Generis
PUBLISHING

AF582370

Object Counting Device Based on IR Technology

Maksims Kotjuks

Title: **Object Counting Device Based on IR Technology**

ISBN: 979-8-89248-563-0

Author: Maksims Kotjuks

Cover image: www.pixabay.com

Publisher: Generis Publishing
Online orders: www.generis-publishing.com
Contact email: info@generis-publishing.com

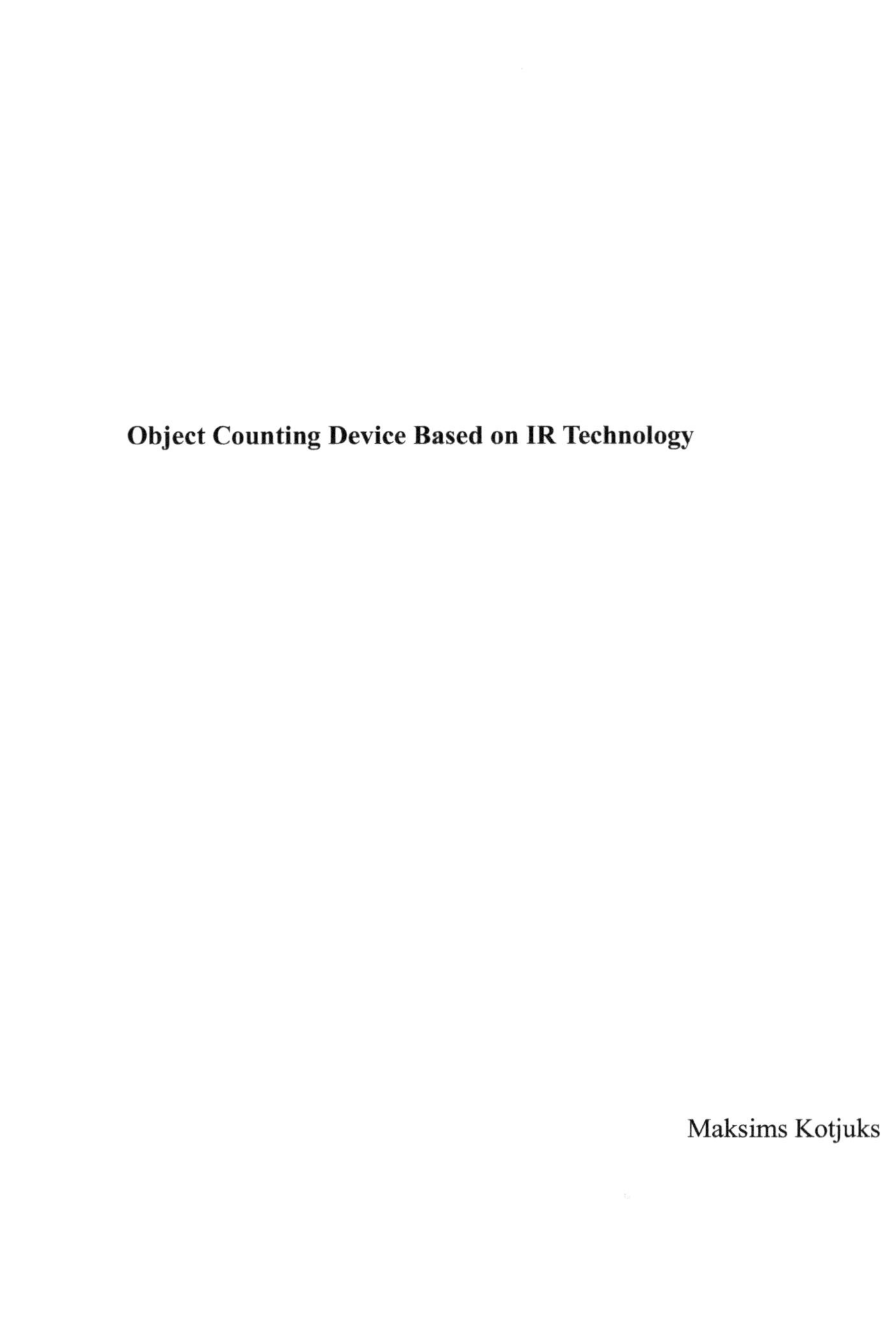

Object Counting Device Based on IR Technology

Maksims Kotjuks

Introduction

At the time of writing this book, I am just an ordinary student. Once I needed to find an electrical circuit diagram, solder and test it for operability, and write a research paper about it for my term project. That's when I found a diagram for an interesting device - an object counter.

An object counter is a device that uses sensors to count items (mainly on conveyor belts), displaying the number on a digital scoreboard or computer program. This book describes the operation of a specific selected device, its schematic, development, and error correction. The research also explores how this device works and what its operational limits are. Additionally, there is a list of components needed to supply, build, and solder to the circuit board.

The book is divided into several main sections: the principle of operation of the device, the implementation of the project, modifications to the device, and calculations and measurements. Additionally, there are supplementary chapters that detail specific components and their functions. Looking ahead, I believe this book can help beginners start their journey in electronics. I have tried to describe everything in detail using simple words so that the reader can understand the basics and acquire some new skills.

This book is an expanded account of my research. Essentially, it is a scientific article featuring a moderate amount of terminology, research, and experiments. The research was conducted with the help of external sources of information, which greatly helped me understand the principle of the device. I hope these sources will also benefit the readers. The references used are provided at the end of the book. Additionally, an appendix is included, where you can find images that did not fit on the main pages of the book.

Table of Contents

List of Abbreviations and Terms

№	Abbreviation	Explanation of abbreviation
1.	IC	Integrated Circuit, a component that contains a circuit composed of other components
2.	IR	Infrared, spectrum of light
3.	TR	Name of the one of NE555 pins - trigger
4.	LED	Light-emitting diode
5.	R	Resistor, symbol in circuit diagram
6.	D	Diode, symbol in circuit diagram
7.	U	IC, symbol in circuit diagram
8.	RV	Potentiometer, symbol in circuit diagram
9.	PCB	Printed circuit board

Principle of Device Operation

The full circuit diagram can be seen on page 29-30.

The chosen circuit can be divided into five important parts (Fig. 1.): a pair of infrared (further IR) diodes (transmitting and receiving diode), a monostable multivibrator, counters and two 7-segment displays. There is a short description of principle of operation:

A pair of IR LEDs is used to detect the presence of an object. The infrared transmitter emits light; if an object is in the path of the emitted light, then the beam is reflected and the receiver, having received the light, sends a signal to the comparator to determine the presence of the object. If it was, then the comparator sends a signal to the monostable multivibrator. The monostable vibrator generates a single pulse, which is fed to the counters input, which, upon receiving this information, displays the current counter value with the help of the 7-segment displays.

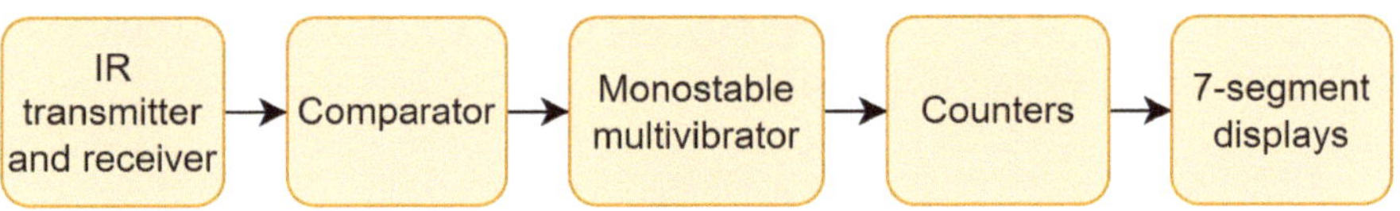

Fig.1. Block diagram of the device

Now, let's explore the working principle in more depth to understand the role of each component.

The electrical principle diagram of the receiver and comparator part is shown in figure 2. The IR transmitter consists of an infrared LED D1 and a current-limiting resistor R1 (setting the current around ~6.7 mA). The receiver part consists of a photodiode D2 and a limiting resistor R2. The signal from photodiode D2 enters the non-inverting input (V+) of the comparator. The comparator is built on the basis of the operational amplifier LM358 (U1A in the diagram). The potentiometer RV1 sets the support voltage of the inverting (V-) input of the comparator.

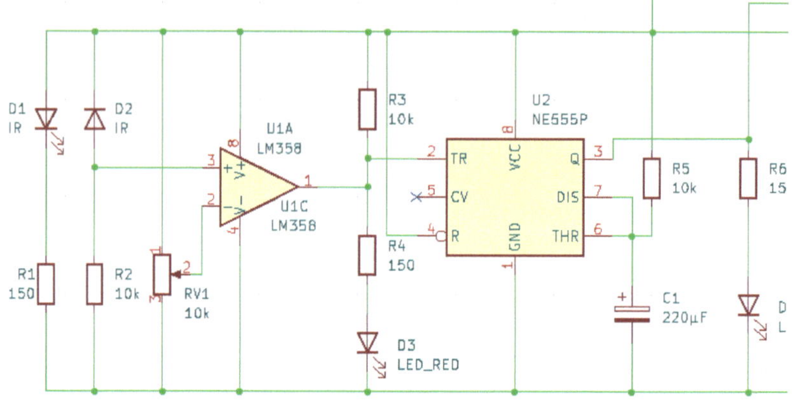

Fig.2. IR diode pair and comparator part

Comparator (lat. comparare - to compare) - a comparison device - a logical electronic device with two inputs and one output. The comparator produces a high voltage at the output (logic 1) if the voltage at the first (direct) input is higher than the second (inverting), and a low output voltage (logic 0) if the voltage from the first input is lower than the voltage from the second. Simply put, a comparator compares two signals and produces a specific result.

To generate the necessary pulses for the counters CD4033, a monostable vibrator is used, created by using a timer NE555 (U2). A monostable multivibrator, as the name implies, has only one stable state. When one transistor conducts, the other remains non-conducting. A monostable multivibrator generates a fixed time period pulse, the duration of which can be set using the given formula:

$$T = 1.1 * RC,$$

where R is the resistance of resistor R5 – 10 kΩ, and C is the capacitance of capacitor C1 – 220 µF. Using this formula, we get a result of 2.42 seconds. This means that the device can read one object every 2.42 seconds.

LED D3 (with current limiting resistor R4 to avoid overpowering) is used as a visual indicator that there is a high signal level at the output of the comparator, indicating that an object has been detected. R3 is used as a pull-up resistor to ensure that the NE555 timer input TR maintains a constant high level when no obstacle is detected.

LED D4 together with the current limiting resistor (R6) serves as a visual indicator that an object is in the sensor's field of view. This does not always indicate that the object is being counted.

The data output part consists of two CD4033 counters/decoders and, respectively, two 7-segment indicators, which have a common cathode (Fig. 3.).

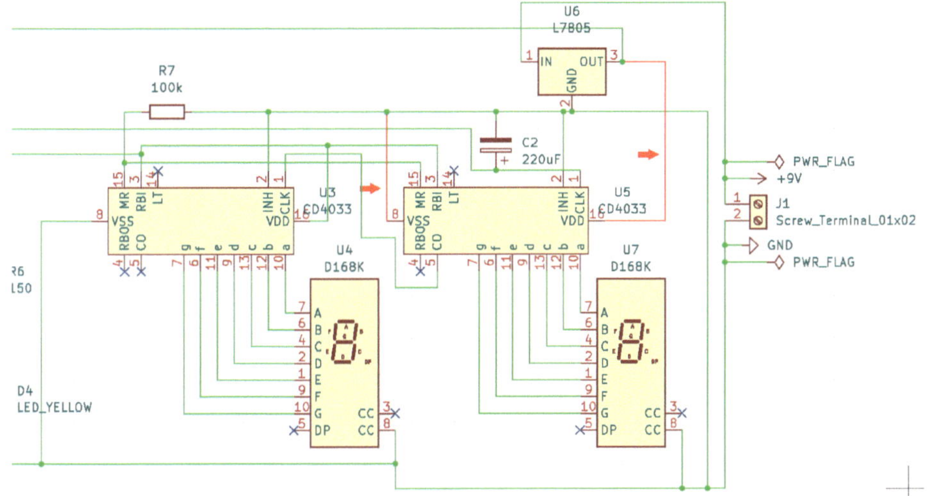

Fig.3. Counter/decoder and indicator part

CD4033 are Johnson counter integrated circuits commonly used in digital indicators. It converts the Johnson code into a 7-segment decoded output. This means that it will convert the input into a digital display that can be seen on a 7-segment display.

Johnson code is a number system in which each subsequent element differs by only one binary digit. The principle of operation of the Johnson code can be seen in the following table:

Number	Johnson code
0	00000
1	00001
2	00011
3	00111
4	01111
5	11111

There is a RESET pin (15th) on the CD4033 IC. Its function is to reset the count completely if the received logic state is high. This pin is not needed in this project; however, according to the manufacturer's recommendations, it should not be left open. Therefore, a high-resistance resistor R7 is used to keep the current very low.

The L7805 voltage regulator is used to decrease the input voltage to 5 volts and stabilize it. The circuit is powered by 9 volts, so the circuit diagram shows a terminal block to which the power supply can be connected.

Characteristics of Integrated Circuits

Operational amplifier LM358

An operational amplifier, commonly known as an op-amp, is an integrated circuit that amplifies the voltage difference between its two input terminals. Op-amps are widely used in electronics for tasks such as signal amplification, filtering, and mathematical operations like addition, subtraction, integration, and differentiation. They have a high input impedance, low output impedance, and are designed to provide a large gain, making them versatile components in analog circuits.

This particular circuit uses the LM358 op-amp. This circuit consist of two independent, high-gain op-amps which are designed specifically to operate from a single power supply over a wide range of voltages (3 – 30 V). The low power supply drain is independent of the magnitude of the power supply voltage.

Pin connections
(Top view)

1 - Output 1
2 - Inverting input
3 - Non-inverting input
4 - $V_{CC}-$
5 - Non-inverting input 2
6 - Inverting input 2
7 - Output 2
8 - $V_{CC}+$

However, as mentioned earlier, the operational amplifier is not only used to amplify signals but also functions as a comparator. It operates as follows: two signals are compared – one at the inverting input and one at the non-inverting input. As long as the voltage at the inverting input is higher than that at the non-inverting input, the output will be negative. If the voltage at the non-inverting input becomes higher than at the inverting input, the output signal will turn positive. In this circuit, a constant voltage is applied to the inverting input of the amplifier through a potentiometer. When an object enters the device's field of view, a photodiode connected to the non-inverting input sends a signal. For the object to be detected, the signal at the non-inverting input must be higher, so the circuit uses a potentiometer to fine-tune the voltage supply.

Timer NE555

The NE555 is a highly versatile and widely used timer integrated circuit. It can operate in various modes, including astable, monostable, and bistable, making it suitable for a wide range of timing and pulse generation applications. In astable mode, the NE555 functions as an oscillator, producing a continuous square wave output. In monostable mode, it generates a single pulse of a specific duration in response to an external trigger. The NE555 is known for its reliability, ease of use, and ability to drive significant current, making it a staple in both hobbyist and professional electronics projects. The operating voltage of the IC is 4.5 - 16 V.

Specifically in this circuit, the NE555 functions as a monostable multivibrator, sending out a signal for a specific time period. This time period depends on the values of the resistor and capacitor connected to the timer as follows:

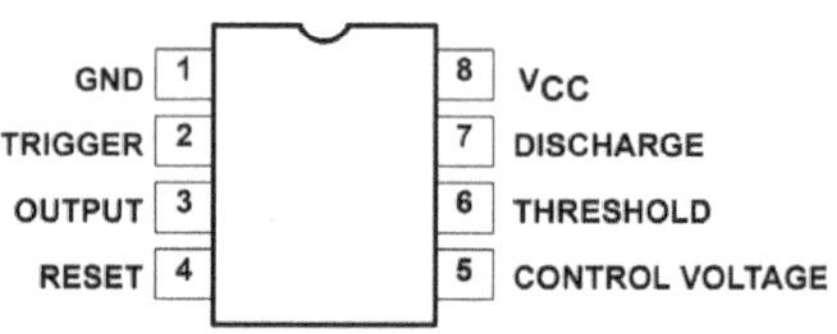

- timer pins 6 and 7 are connected to each other,
- a resistor is connected between pin 6 and power,
- a capacitor is connected between pin 7 and ground.

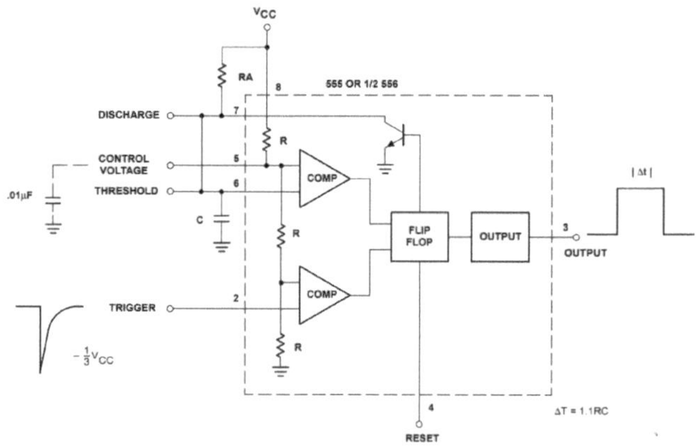

Fig. 4. The principle of operation of the timer in monostable mode (taken from the official NE555 datasheet)

Decade Counter CD4033

The CD4033 is a decade counter IC that integrates both counting and display-driving capabilities, making it a popular choice for digital counting applications. It counts from 0 to 9 and directly drives a 7-segment display, eliminating the need for an external decoder. The chip features a built-in 7-segment decoder/driver, which converts the binary count into the corresponding display pattern, making it easier to implement visual counting systems like digital clocks, event counters, and other numeric displays.

One of the key features of the CD4033 is its ability to be cascaded with other CD4033 chips to count beyond 9, which is particularly useful in multi-digit displays. For instance, by connecting several CD4033s in series, you can create a counter that displays numbers in the tens, hundreds, or thousands range. The **CARRY-OUT** pin allows the overflow from one counter to increment the next, ensuring seamless counting across multiple digits.

Top View

CLOCK 1 — 16 V_{DD}
CLOCK INHIBIT 2 — 15 RESET
RIPPLE BLANKING IN 3 — 14 LAMP TEST
RIPPLE BLANKING OUT 4 — 13 c
CARRY OUT 5 — 12 b
f 6 — 11 e
g 7 — 10 a
V_{SS} 8 — 9 d

92CS-24475RI

CD4033B

In terms of usage, the CD4033 is favoured for its simplicity and versatility in both hobbyist and professional projects. It operates on a wide range of supply voltages (from 3V to 15V), making it compatible with various circuits. Its low power consumption and ability to drive LEDs directly make it an efficient and cost-effective solution for creating digital counters, scoreboards, timers, and frequency meters. The straightforward pin configuration and minimal external components required further enhance its appeal for those looking to build reliable and easy-to-use counting circuits.

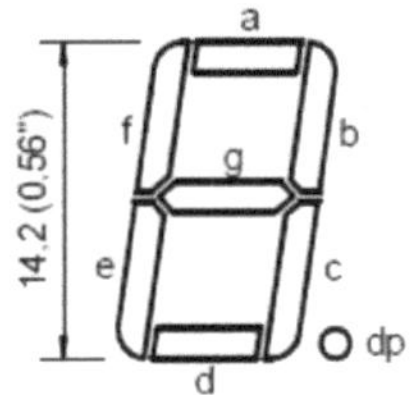

Pins 6-7 and 9-13 on the CD4033 are responsible for controlling specific segments of the 7-segment display. Each of these pins must be correctly connected to the corresponding inputs of the display to ensure that the correct segments light up and the intended number is displayed. Therefore, it is essential to check which pin on the 7-segment indicator controls each segment before making connections.

Diagrama Elétrico Interno

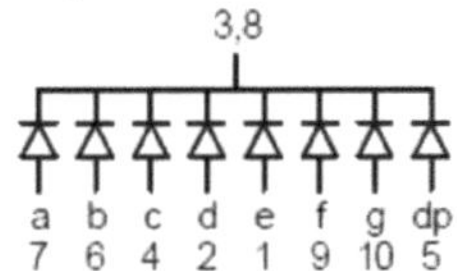

The CD4033 has a **LAMP TEST** input which, when connected to a high-level voltage, overrides normal decoder operation and enables a check to be made on possible display malfunctions by putting the seven outputs in the high state.

RBI is used to suppress leading zeros on a 7-segment display. When this input is active, it blocks the display of all zeros up to the first significant digit. **RBO** controls the suppression of zeros on subsequent digits, which is important for multi-digit displays. It allows the transmission of information about whether it is necessary to suppress zeros on the next digit. For example, if multiple displays are used and it is necessary to remove excess zeros (e.g., when displaying 0042.7800), you should connect it as follows:

- RBI on the first counter to ground,
- RBO on the first counter to RBI to the second counter,
- Repeat connecting RBO to RBI for all subsequent counters,
- The last counter in the cascade should have RBI connected to ground.

There is a closely related analogue of the CD4033: the CD4026. The primary differences between the two are that the CD4026 features a **DISPLAY ENABLE** pin and an **UNGATED "C" SEGMENT** instead of the **RIPPLE BLANKING** and **LAMP TEST** pins found on the CD4033.

- **DISPLAY ENABLE** is used to reduce the power consumption of the circuit when the display is not in use.
- **UNGATED "C" SEGMENT** is employed when the display needs to show numbers divided by 12 or 60, making it suitable for applications such as digital clocks.

Project Implementation

Having understood the principle of operation of the object counter, let's begin testing the circuit and soldering it. Here is a complete list of components to implement the device (the number in brackets indicates the quantity):

- resistors: 150Ω (3), 10kΩ (4), 100kΩ;
- electrolytic capacitors 220μF (2);
- LEDs: red, yellow, IR;
- IR photodiode;
- NE555 timer;
- LM358 amplifier;
- CD4033 (or CD4026) Johnson counters (2);
- 7-segment indicators (2);
- L7805 voltage regulator;
- screw terminal block.

It is not for nothing that I mentioned testing the circuit - this will help in case the original circuit is faulty. To avoid soldering and wasting components, a breadboard can be used. It is advisable to use a laboratory power supply as it provides a constant and stable power source and protects against short circuits and high voltage surges in the circuit.

Once the components are placed on the board and connected, the circuit can be connected to the power supply. However, after connecting the circuit, I noticed that one of the indicators did not light up.

Fig. 5. As seen in the photo, only one indicator is turned on

In such situations, it's important to double-check that all components are connected correctly. After ensuring that the indicator was connected correctly and verifying the counters' connections, I began to look for a problem in the circuit itself. Based on the issue, there could be three possible causes: a malfunctioning indicator, a malfunctioning IC, or incorrect connection of components in the original circuit diagram.

It is easy to check the performance of the indicators - just connect them to a 2V power supply. In 7-segment indicators, the cathodes (negative pins) are usually pins 3 and 8, and the rest are anodes (positive pins); each segment will light up separately. Checking the IC is more difficult, as it is a logical element with many inputs and outputs that function under specific conditions. I decided to skip this step and move on to the third reason - an error in the circuit diagram. I paid special attention to the non-working indicator and the IC CD4026 connected to it. I reviewed the datasheet of the CD4033/CD4026 (they are equivalent) and noticed a detail in the connection of these ICs: the IC with the lit indicator was connected to the power supply, but the second IC was not. This issue has a simple solution - just connect the second chip to the power supply. After doing so, the indicator lit up, and testing of the device could begin.

Using a potentiometer, the device needs to be calibrated so that it correctly detects and reads the object. During calibration, I noticed that either the yellow or red LED lights up depending on the direction in which the potentiometer is turned. For the object to be properly detected, the potentiometer should be adjusted to a position where the LEDs are just about to switch between each other, ensuring that only the yellow LED lights up.

After setting up the device, I began testing its ability to read objects. When an object entered the field of view of the IR beam, the red LED lit up while the yellow LED turned off. The red LED consistently indicated the presence of an object within the device's field of view, whereas the yellow LED turned off only when the object was read. As mentioned earlier, the object counter is designed to have a delay between reading successive objects, which, according to calculations, should be approximately 2 seconds. Although the object counter accurately counted the number of objects passing its field of view, I observed another issue: the device frequently skipped 3-4 counts. I started investigating to identify the source of the problem.

The first step I took was to replace the potentiometer with a less sensitive one. Potentiometers with a knob are quite sensitive and can shift their position due to slight vibrations or accidental touches. Since the object counter requires precise adjustment, I opted for a different type of potentiometer known as a trimmer potentiometer. This type requires adjustment with a screwdriver, which minimizes the risk of accidental changes.

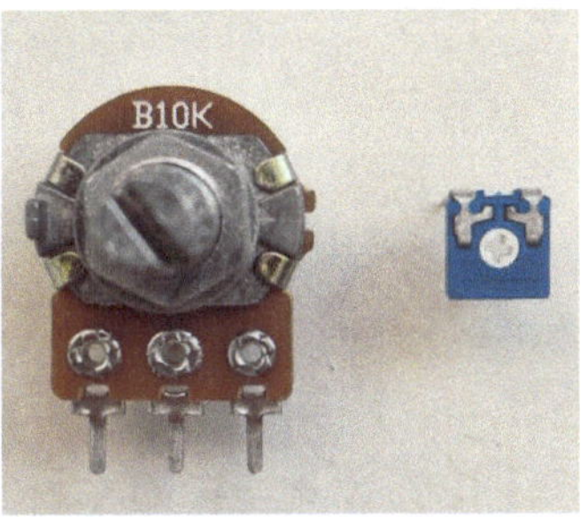

Replacing the potentiometer did not resolve the issue, so I proceeded to the next step: checking the signals in the circuit. I hypothesized that the object counter was jumping values because it was receiving multiple signals in a very short period of time, causing the device to read several objects instead of just one. I used an oscilloscope to visually assess the problem.

First, I connected the oscilloscope to the junction of the photodiode (anode) and the LM358 amplifier (pin 3); the signal was stable when an object was detected. Next, I connected the oscilloscope to the junction of LM358 (pin 1) and NE555 (pin 2); again, the signal was stable, ruling out this part of the circuit as the issue.

Finally, I connected the oscilloscope to the junction of NE555 (pin 3) and CD4026 (pin 1). Here, my hypothesis was confirmed: the counter was indeed receiving multiple signals in a very short time when reading a single object (Fig. 5). To address this, I connected an electrolytic capacitor with a value of 220 μF between pin 3 of the NE555 and ground to smooth out the multiple short signals into one. The results were immediate—the counter began to operate correctly, allowing me to move on to the next stage of development: soldering the components.

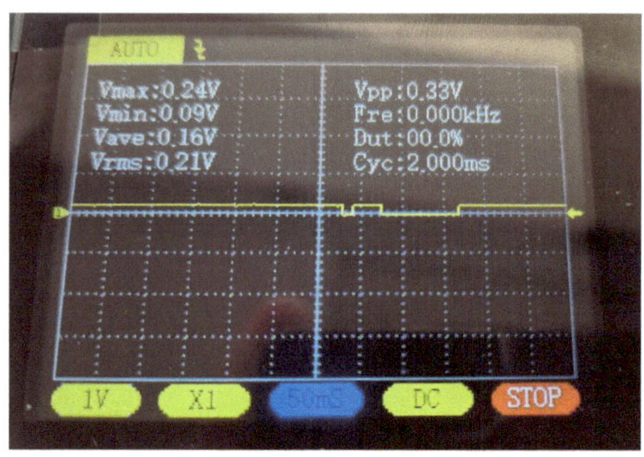

Fig. 6. This is what the unstable signal looked like: several pulses were received in a short period of time instead of just one

Fig. 7. Working prototype of the object counter

Before soldering the circuit, it is important to choose the type of board to which the components will be soldered. There are several options available, you can use:

- a solderable breadboard; in this case, you will need to use wires to connect the components to each other;
- a copper-clad laminate (CCL); this requires drawing paths, drilling holes, and etching the board;
- a computer software that helps you to draw a circuit diagram, design a circuit board and then order from PCB manufacturers; this option involves additional costs for board production and shipping, however, this approach helps save time during soldering, as the components are already connected on the board; all that remains is to solder the components themselves to the PCB.

I decided to implement the project using the third option. I used KiCad software to design the PCB - it's one of the free programs with a large library of component layouts and is easy to use. The steps for creating the board were as follows: draw the schematic → assign layouts for components → place the components on the virtual board and connect them with tracks according to the schematic → export the production files and place the order with the manufacturer. I used a track width of at least 0.5 mm to prevent the board from heating up due to current flow (narrower tracks have higher resistance, leading to heat generation). I also added a screw terminal block to the schematic to supply the device with power and some mounting holes. The result of completing all the steps can be seen in the photo below.

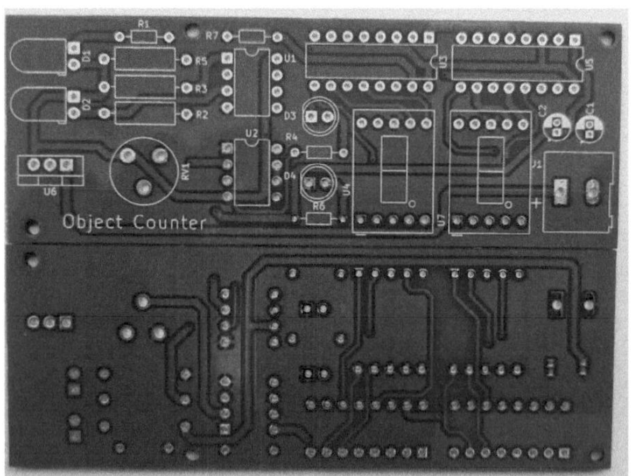

Fig. 8. Printed circuit boards

After receiving the board, it was necessary to solder the components onto it. After testing the board for functionality, additional tests could be conducted.

<u>Fig. 9. This is what the working board looks like. The photo shows that when an object is within the IR beam's field of view, the LEDs change state as described earlier. When the object moves out of the device's field of view, the number will change to 38.</u>

Testing the device

To determine how the object counter behaves in certain situations, some research is needed. The technical characteristics of the device can be evaluated by considering the following questions:

- How much power does the device consume?
- What is the minimum and maximum distance at which the device can detect objects?
- Will all types of surfaces reflect light well enough for the meter to count an object?

To determine how much power a device consumes, a simple formula from basic physics can be used: $P = U \cdot I$ (P – power, watts; U – voltage, volts; I – current, amperes).

We know that the object counter consumes 9 V, but to determine the current consumption, a multimeter needs to be connected in series between the power supply and the device. It is worth noting that the counter device has a 7-segment display, which means that the current consumption may vary depending on the number being displayed. Therefore, statistical formulas will be used to account for this variation. To determine the average power consumption of the device, the arithmetic mean formula was chosen (the sum of all values divided by their number); looking ahead, I will say that this was the best option for calculation, as the data showed little variation and did not stand out overall. To calculate the deviation from the average value, the standard deviation formula was selected, which is as follows:

$$s = \sqrt{\frac{\sum_{i=1}^{n}(x_i-\bar{x})^2}{n-1}},$$

where:

- x_i – individual value,
- $\bar{x}$ – arithmetic mean value,
- n – total number of values.

Of course, third-party programs (such as MS Excel) or online resources can be used for calculations – this makes it easier to handle many measurements.

Speaking about the object counter, after conducting the experiment and performing several calculations, it can be noted that the device's energy consumption generally varies between **0.6** and **0.85 W** (with only 3 values falling outside this range). The average consumption is **0.736 W**, with a standard deviation from the mean of **0.071 W**. It is worth mentioning that measuring devices also have their own margin of error. The multimeter used in the experiment has an error of ±2% + 0.02 mA, which is significantly small and can be omitted from the overall calculations. We can proceed to the next experiment.

I decided to first conduct an experiment to determine which surfaces the device can read best. The next experiment – measuring how far the counter can detect an object – will depend on these results.

First, I tested how far the device could detect an object with a glossy surface. The distance was measured from the end of the LED using a ruler. The maximum distance at which the object counter could detect an object with a glossy surface was 80 mm, with the optimal distance being 75 mm. For an object with a matte surface, it was detected up to 60 mm, but the detection was inconsistent, making the optimal distance for such a surface 45 mm. Metal objects performed very well: the counter could detect them at a distance of up to 180 mm, provided the surface was smooth and the reflective beam was directed straight towards the photodiode. The device could also detect transparent objects, but the distance was no greater than that for matte objects. It is worth noting that colour also played a role: a matte black object (black fabric was used during the test) was not detected by the counter at all, regardless of how close the object was.

During this test, it was found that surface unevenness can also affect the detection distance. For optimal detection, the surface of the object should be smooth and oriented perpendicular to the photodiode. This issue might be due to the narrow operational range of the IR LED or photodiode, requiring the IR beam to be precisely directed at the photodiode. A potential solution to this problem could be installing a lens on the photodiode, which would provide it with a wider viewing angle.

Technical Characteristics of the Object Counter

Technology used	Logic elements and IR light
Input voltage	5 – 35 V*
Power consumption	~0,74 ± 0,07 W
Reading speed	Approx. 24 objects/min
Readable objects and max distance for optimal working	• Metal – 170 mm, • Glossy – 75 mm, • Matte – 45 mm, • Transparent – 45 mm, **Black matte objects are not readable**

* The input voltage varies because the entire circuit is powered by a voltage regulator L7805. According to the manufacturer, this regulator can be supplied with voltage up to 35 V.

Ways to Improve the Device

Here, I will provide suggestions for improving and modifying the object counter. The reader can experiment with these ideas to further refine the device, customizing it to their preferences and potentially bringing it to perfection.

- Consider using fixed resistors instead of a potentiometer. This would eliminate the need for the owner to adjust the device manually. However, determining the ideal resistors for this situation may require considerable experimentation.
- On the forum discussing the original circuit diagram, it was mentioned that the number of indicators can be increased, allowing the count to extend to 999, 9999, or even 99999. Review the circuit again and try to expand the number of readings according to your requirements and preferences. Additional details can be found in this book (see page 26).
- When connecting a device directly to the power supply, interference may sometimes occur at the moment of connection, potentially causing malfunctions in the device. To prevent these issues and to make turning the device on and off easier, a switch can be added to the circuit between the power supply and the rest of circuit.
- Since the entire circuit is powered by 5V via the voltage regulator, you might consider removing this component and using a different connection port, such as USB Micro or Type-C, instead of the terminal block. This way, you can connect the circuit directly to a power outlet, eliminating the need for batteries. Alternatively, you could assemble your own power supply for the device.
- Try using lenses, as mentioned earlier. This could help increase both the detection radius and reading distance. It is important to use a converging lens and attach it to the photodiode. Due to the likely size of the lens, you may need to adjust the positions of the IR LED and photodiode. Instead of keeping them parallel as in the original circuit, try positioning them at an angle and directing them toward each other.

Oscilloscope Measurements

To visually observe how the object counter operates, an oscilloscope was used. The oscilloscope was connected at three points: pin 3 and pin 1 of the LM358, and pin 3 of the NE555.

First, the oscilloscope was connected to pin 3 of the LM358 to observe how the signal from the photodiode was fed to the comparator. When an object entered the counter's field of view, the photodiode generated a small voltage signal. The duration of the signal depended on how long the object remained within the counter's field of view.

Next, after the comparator, the signal was sent to the NE555 timer, so the oscilloscope was connected to pin 1 of the LM358. The signal duration also depended on how long the object remained in the device's field of view. It is important to note that the signal coming out of the comparator was amplified, as the LM358 is an operational amplifier. This amplified signal also drives the red LED, causing it to light up when the counter detects an object.

After the timer, the signal is sent to the CD4026 counters, so the final point where the oscilloscope was connected was pin 3 of the NE555. At this point, you can observe that the signal is inverted, meaning that when an object is detected, the voltage drops. This signal is only present if the object has been successfully read, not just if it is visible to the device. This signal also controls the yellow LED; while the device is not reading an object, the LED remains on, but as soon as the object is counted, the LED turns off. The voltage change impacts the CD4026 counter, causing it to update the digits on the 7-segment display as soon as the object leaves the device's field of view.

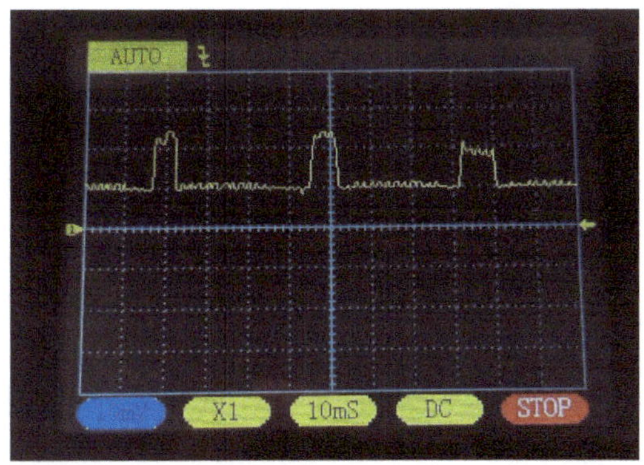

Fig. 10. The oscilloscope was connected to pin 3 of the LM358

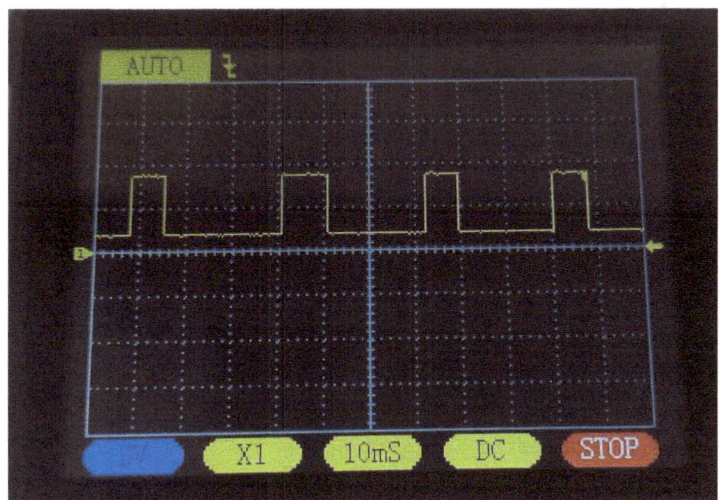

Fig. 11. The oscilloscope was connected to pin 1 of the LM358

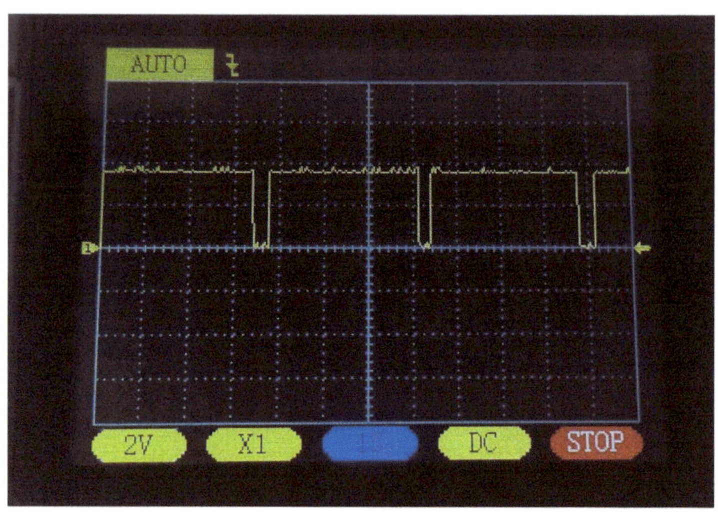

Fig. 12. The oscilloscope was connected to pin 3 of the NE555. You can also observe that one object was counted approximately every 2.5 seconds

Application of the Device

The object counter can be used in a variety of applications, including:

- Industrial manufacturing: to count parts or finished goods on production lines, helping to track production output and manage the production process.
- Retail: to count customers entering a store or items passing through a checkout, aiding in the analysis of customer traffic and sales trends.
- Access control systems: in areas with restricted access, counters can be used to track the number of people entering or leaving the premises, which is crucial for maintaining security.
- Transportation: in public transportation, to count passengers, which assists in optimizing routes and scheduling.
- Automation and robotics: to monitor the number of objects picked up or moved by robotic systems, ensuring the accuracy of task execution.
- Farming: to count fruits, vegetables, or other products on packing lines, aiding in crop management and processing.

It is worth noting that this object counter may be challenging to use in applications 3, 4, and 5, as not all objects will be accurately detected by the device, as mentioned earlier.

Improving the Object Counter

In this chapter, I will discuss how to improve the device and adapt it for different purposes.

First, let's examine which parts of the device can be improved. One of the most obvious modifications is increasing the number of 7-segment indicators, allowing the device to count a larger number of items. Enhancing this aspect of the counter is not as challenging as it may seem. Let's take a closer look at the circuit diagram.

As you can see, each 7-segment indicator is paired with its own CD4033 counter. Therefore, to increase the number of indicators, you need to add more CD4033 ICs. For example, if you want to extend the count to 9999 objects, you'll need 4 indicators and 4 counters – one for each indicator. However, this raises the question: how should they be connected properly?

The way the counters and indicators are connected does not change – each CD4033 output corresponds to a segment on the indicator. However, for the ICs to interact with each other and the rest of the circuit, they must be connected as follows:

- All pins '15' on the counters should be linked together and connected to a 100 kΩ resistor.
- All pins '3' and '16' should be connected to each other and then to the output of the L7805 voltage regulator.
- All pins '2' and '8' should be connected to each other and to the negative power output (ground).
- Regardless of the number of counters used in the circuit, it is important to properly connect the last (the largest unit) and first (the smallest unit) indicators. The counter connected to the first indicator should be wired as follows: pin '1' of the counter to pin '3' of the NE555, and pin '5' of the counter to pin '1' of the next counter, as in the original circuit. For all intermediate counters, connect pin '5' of the counter of the smaller unit to pin '1' of the counter of the next unit. Continue this sequence until you reach the last indicator, where pin 5 should remain unconnected. The operating principle of the counters was described in chapter "**Characteristics of Integrated Circuits**"

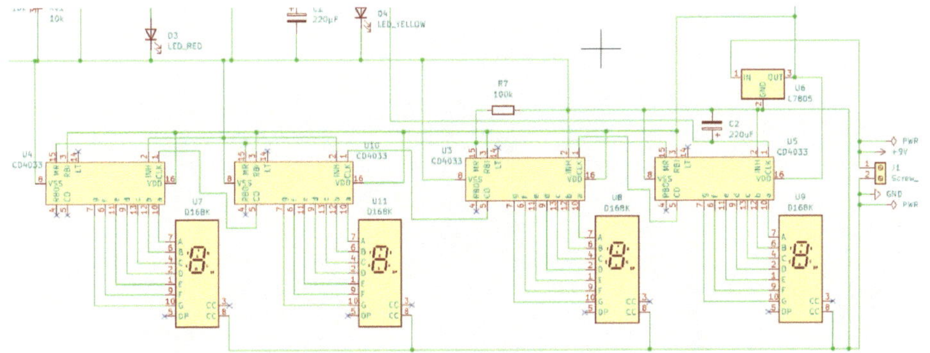

Fig. 13. Visual example of connecting four indicators

To modify the device, the reading speed can also be increased. As noted earlier, the components responsible for the reading speed are R5 and C1. According to the formula, to achieve a faster reading speed (i.e., to decrease the time period between readings), the value of one or both of these components should be reduced.

Before assembling the final circuit, it is important to verify that these modifications are effective and do not disrupt the device's operation. Therefore, the circuit was reassembled on a breadboard. Fortunately, the object counter functioned correctly without any issues, so no further adjustments were necessary. The next step is to proceed with the soldering stage.

Fig 14. Prototype of a modified working object counter.

I decided to follow the same approach as with the original circuit assembly – design the circuit in a computer program and order a PCB from suppliers. However, due to the modifications, the number of required components increased. As a result, the size of the board also grew, so in this project, I opted to use SMD (Surface-Mount Device) components to keep the device compact. Components in this form factor are small, taking up minimal space on the board, and allowing components to be placed on both sides of the board. I kept some components the original size, such as the 7-segment indicators, to ensure that the numbers on the display remain clearly visible.

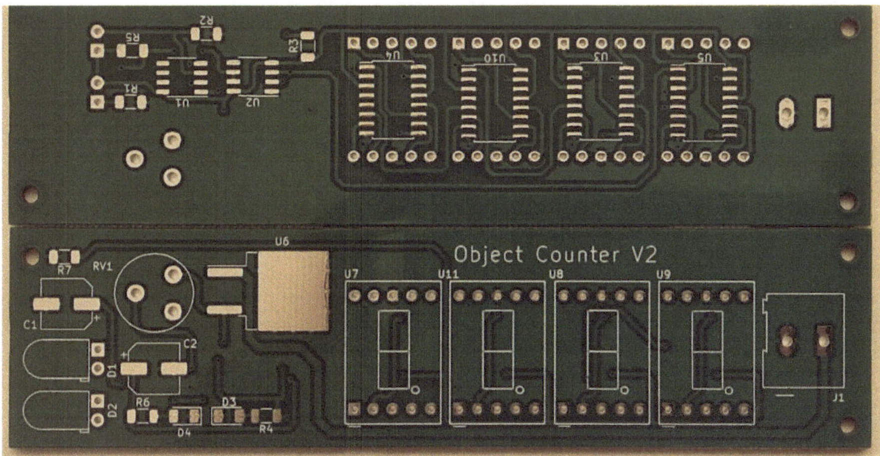

Fig. 15. Printed circuit boards

In this modification, the value of resistor R5 is set to 2.7 kΩ to reduce the reading period for objects, while capacitor C1 remains at its original value. According to calculations, the counter should read an object approximately every 0.65 seconds.

Fig. 16. This is what the working modified board looks like

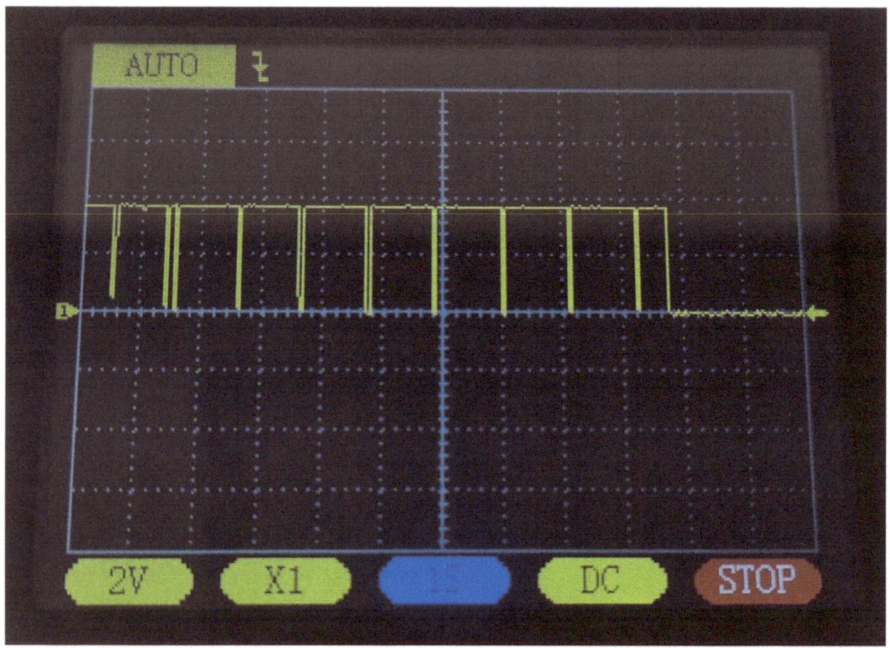

Fig. 17. Shown here are the signals received by the CD4026 counters when an object is read. As calculated, an object is read approx. once every 0.6 seconds.

Technical Characteristics of the Object Counter (Version 2)

Technology used	Logic elements and IR light
Input voltage	5 – 35 V
Power consumption	~1 W*
Reading speed	Approx. 95 objects/min
Readable objects and max distance for optimal working	• Metal – 170 mm, • Glossy – 75 mm, • Matte – 45 mm, • Transparent – 45 mm, **Black matte objects are not readable**

*Since the number of indicators was increased, calculating the exact power consumption would be challenging (requiring measurements for all 9999 readings). Therefore, the most frequently measured current value – 0.11 A – was used as the standard.

Conclusions of the Research

The following conclusions were drawn during the research:

1. The importance of verifying the accuracy of information sources. It's crucial not to design a device without first creating mock-ups or prototypes, as even authors can make mistakes. A poorly designed device may end up failing.
2. Speaking specifically about this object counter, it turned out to be a fairly accessible device that, in my opinion, even a beginner can assemble. The circuit uses readily available and inexpensive components, and the circuit diagram itself is straightforward and doesn't even require programming.
3. This device is highly modifiable, allowing users to customize it to suit their specific needs.
4. Measuring instruments like a multimeter and oscilloscope are essential for characterizing a device.
5. Analysing the circuit diagram and component datasheets can provide a deeper understanding of the principle of the device operation.

Overall, the project was both straightforward and interesting. Through this research, we gained a detailed understanding of the ICs principles, which can be applied to future implementations. We also learned how the device is divided into logical parts and how these parts interact with each other. Most importantly, the device is safe to use. The voltage and current it consumes are low enough not to pose a risk, allowing for safe experimentation. However, I believe the object counter is not yet perfect – there's still room for improvement. This book can serve as a valuable resource for further research.

Appendix

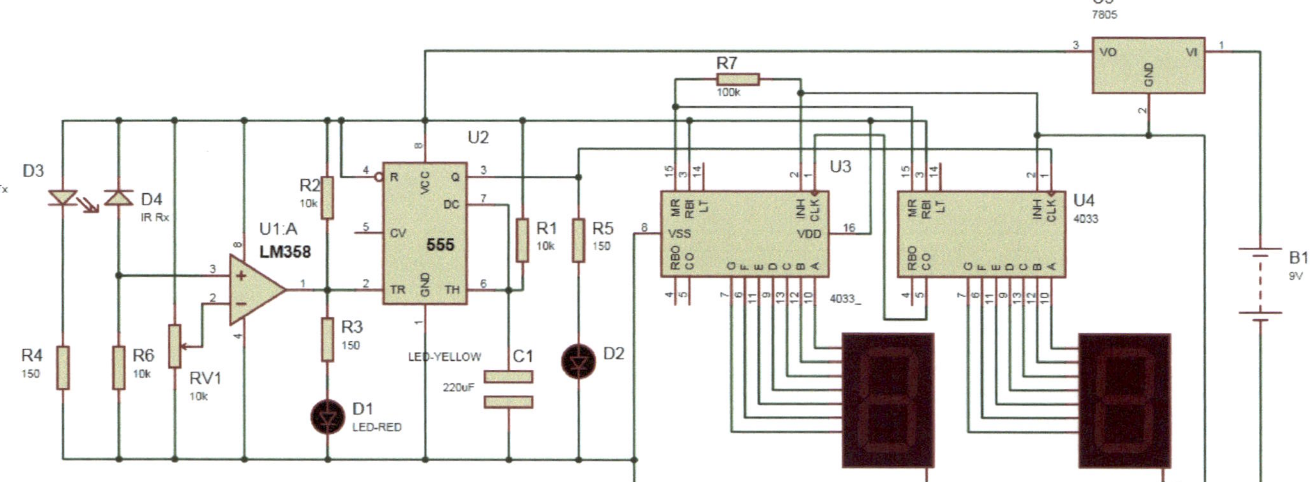

Appendix 1. Original object counter circuit diagram

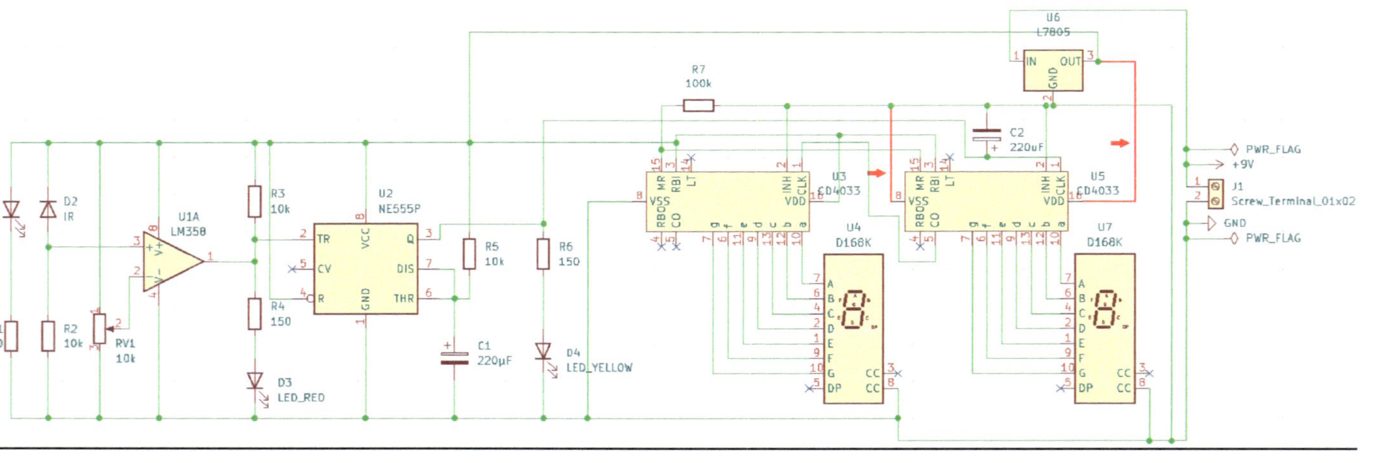

<u>Appendix 2. A slightly modified circuit diagram that works properly. The red arrows indicate the connections that were missing in the original circuit for it to function, and capacitor C2 has been added</u>

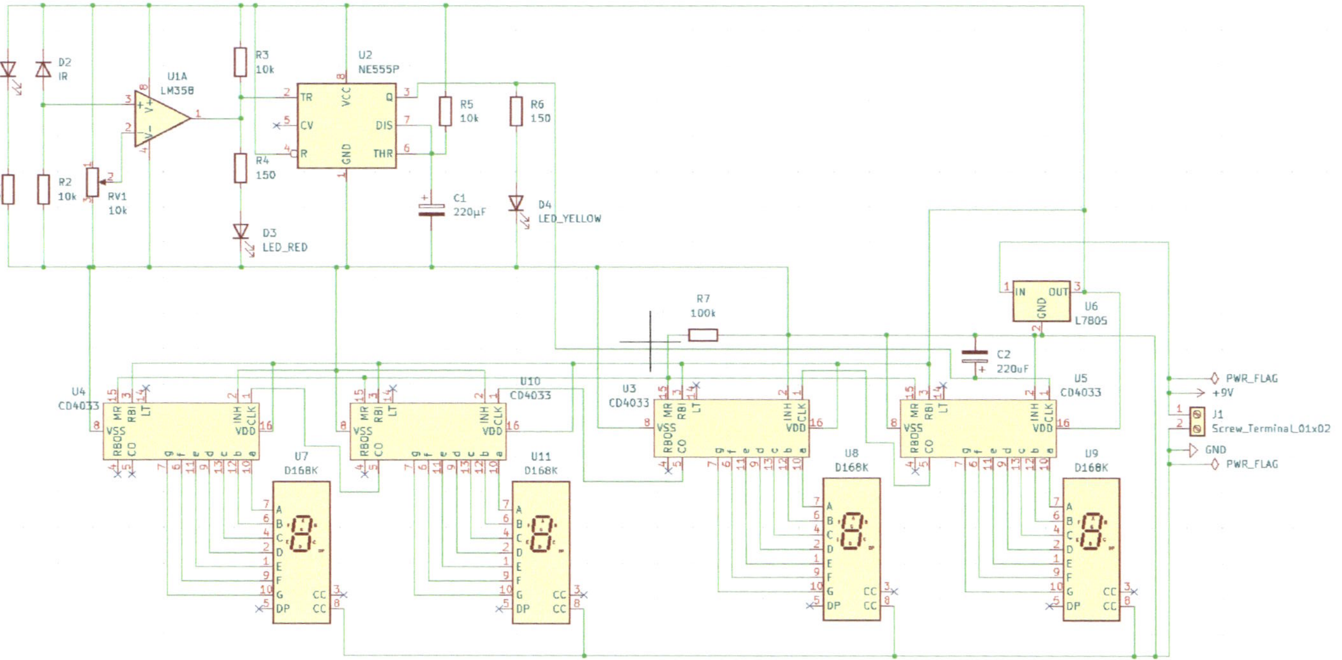

Appendix 3. Circuit diagram with four 7-segment indicators

Sources of Information

This research was conducted with the help of third-party resources. The homepages or titles of the documents from which the information was taken will be listed here.

1. https://circuitdigest.com/ [Original circuit diagram]
2. https://mobilradio.ru/information/ [The principle of operation of the comparator]
3. https://coderlessons.com/ [The working principle of a monostable multivibrator]
4. https://www.engineersgarage.com/ [Operating principle of the IC CD4033]
5. Datasheets: L7805, LM358, NE555, CD4033 (can be freely found on the Internet)

www.ingramcontent.com/pod-product-compliance
Lightning Source LLC
LaVergne TN
LVHW021310160826
845679LV00001B/289
* 9 7 9 8 8 9 2 4 8 5 6 3 0 *